GUM TREE — AN AUSTRALIAN ICON

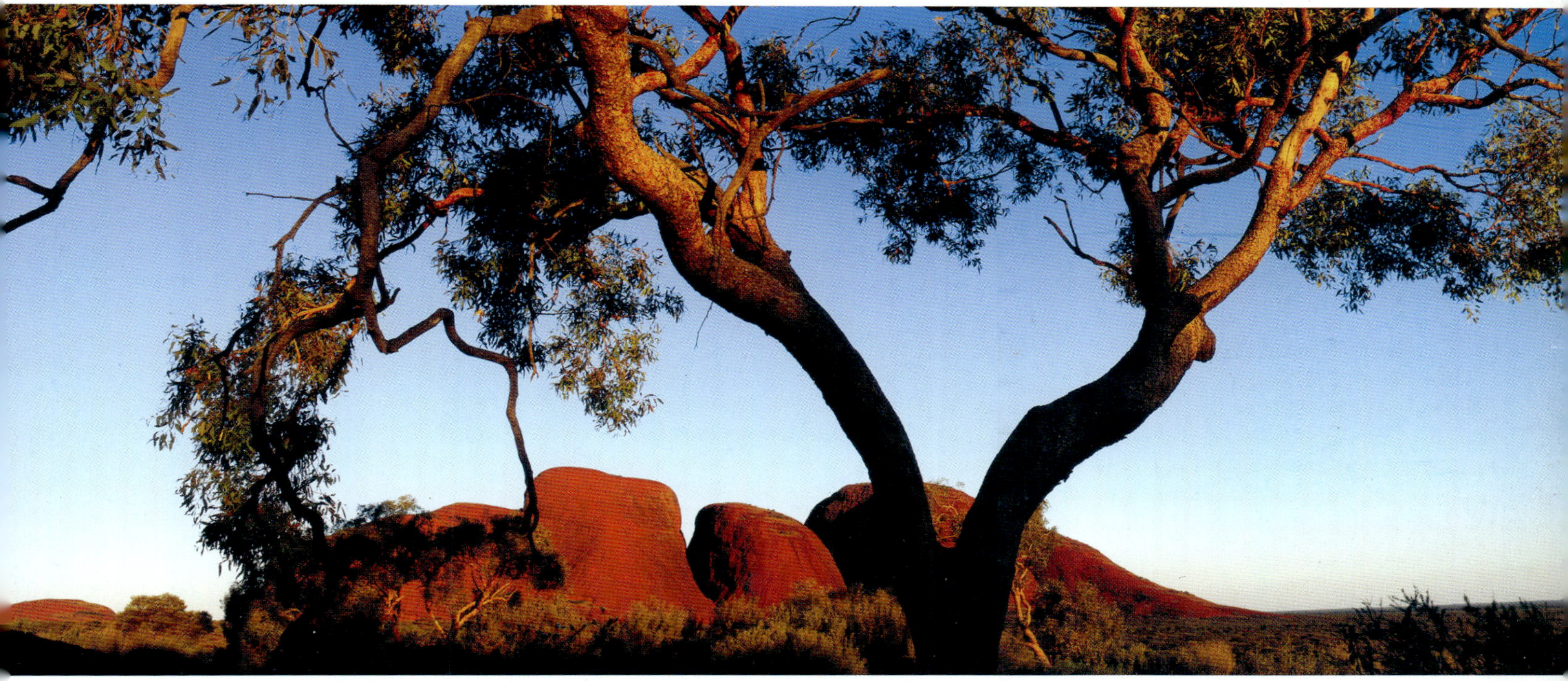

Steve Parish

Gum
Tree

An Australian Icon

Gum Tree

Australia is graced with many unique landmarks (both natural and artificial) that distinguish it from the rest of the world. Our most famous icons – the Sydney Opera House and Uluru – are powerful symbols of the continent, immediately recognisable by overseas visitors. For Australians, however, nothing quite captures the real essence of their homeland more than the humble gum tree. With over 700 species, eucalypts dominate an enormous range of Australian environments – from the arid red centre to freezing sub-alpine highlands. They are versatile and adaptable trees. In so many ways these floral mascots are characteristic of the country they represent – rugged, resilient, and beautifully understated.

This book has been created to celebrate the mighty eucalyptus, the tallest flowering plant on Earth.

Steve Parish

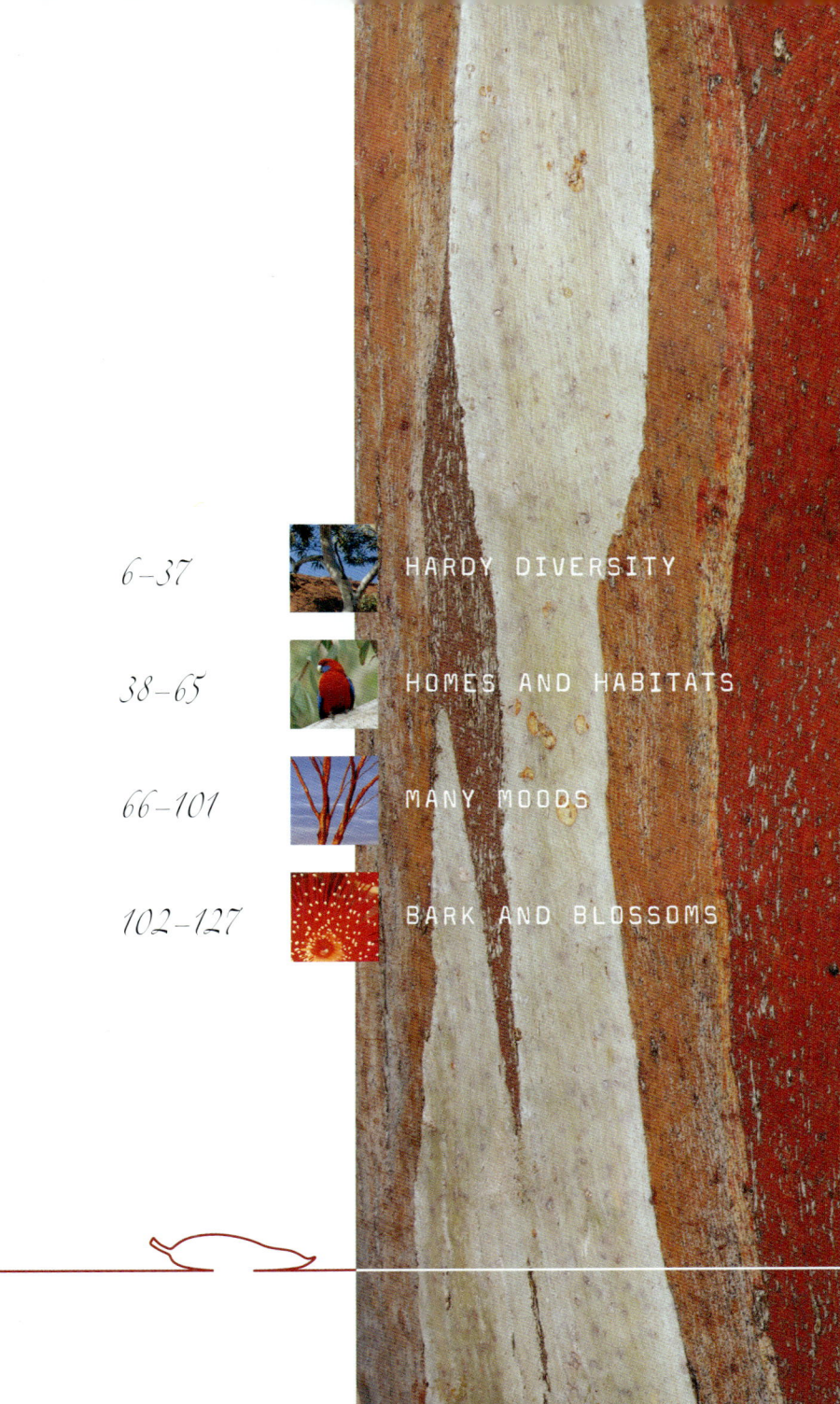

Adapted to almost every Australian environmental condition,

the gum tree is as tenacious as it is varied.

Hardy Diversity

The Blue Mountains, New South Wales, take their name from the bluish tinge to the air – caused by tiny droplets of vapour released from the forests of eucalypts.

Gums cluster around rocky outcrops in the Grampians National Park, Victoria.

Crowning the unforgiving rocky crags,

are scraggly slender branches

stretching

to the sky.

An army of Red Tingle and Karri eucalypts in the forests of Walpole–Nornalup National Park, Western Australia.

Peering skywards through the eucalypt trees of the Valley of the Giants, Western Australia.

Mountain Ash – tallest flowering trees in the world.

Two of Australia's most lauded architectural structures, the Sydney Opera House and Harbour Bridge, are framed by a more simplistic Australian icon – the gum tree.

The surreal shapes of branches and base, warped by the elements, contrast with the delicate beauty of this field in the Flinders Ranges, South Australia.

Landscapes peep from beyond  *gnarled*

branches and peer around *twisted* trunks.

Gums dot the plains but crowd around small water-ways in Queensland's Channel Country.

Australia in abstract – the red bulk of Kata Tju̱ta rises beyond the twisted limbs of gum trees.

Subtle colours and textures of Blue Gum bark.

On the broad banks of the Murray River, South Australia, mighty gums thrive on the rich, alluvial soil and help reduce soil salination.

Lining the rivers and watercourses these *stately* stalwarts quench their age-old thirst...

Stands of eucalypts cover the undulating hills that rise to the bluffs of the Stirling Ranges in south-west Western Australia.

Australia is home to more than 700 species of eucalypt, this one thrives in rocky country in Victoria's north-west.

The Ghost Gum is a silvery vision of floral beauty in Watarrka National Park, Northern Territory.

The colours of the Australian Outback,

its rich russets and faded saffrons, seem to *leach* from the soil

and rock to colour bark and sticky sap...

River Red Gums in the Flinders Ranges, South Australia.

Snow gums, like *graceful* ladies on a winter walk, wrap snow, like shawls, around their elegant frames...

Snow Gums are able to withstand snow and frost down to −20°C in Kosciuszko National Park, New South Wales.

Diamond-bright light sparkles off a hillside in the Snowy Mountains, New South Wales. Most eucalypts are evergreen and retain their leaves in winter.

Gums wait out the winter snows in the unforgiving alpine terrain of Mount Wellington, Hobart, Tasmania.

An avenue of Lemon-scented Gums line Fraser Avenue in Kings Park, Perth, Western Australia.

Indefatigable Mallee eucalypts, like this one in Wyperfeld National Park, Victoria, may survive fire and trauma that levels most other flora.

Horses shelter by a gum in the Mount Lofty Ranges, South Australia.

Overseers of the stock, they *stand*

grey-green in verdant paddocks and pastures...

The rich pastures of the Central New South Wales hinterland support herds of cattle and stands of eucalypts.

Gum trees in the foreground are dwarfed by the imposing blue bulk of Cradle Mountain beyond, Cradle Mountain–Lake St Clair National Park, Tasmania.

A light mist filters between dense forest in the Styx Valley, Tasmania.

In crag and crevice or in the misty heights of
mountain forests, towering gums
endure...

Broad branches, snug hollows, and even waxy leaves provide sanctuary and sustenance for the animals of the Australian bush.

Homes and Habitats

Eucalypts cast long reflections in the glassy water of Cockle Creek, Recherche Bay, Tasmania.

Laughing Kookaburras (top) and bright Crimson Rosellas (bottom) make their homes in the treetops.

Kookaburra sits in the old gum tree... Merry,

merry king of the bush is he...

Laugh kookaburra, *laugh*

kookaburra...

A Brown Falcon surveys the scene from a branch.

The brilliantly plumed Crimson Rosella brings a splash of colour to the artwork of a Scribbly Gum.

The Kookaburra can often be heard laughing raucously from high in the branches.

A Pied Butcherbird grips the flaky branch of a Salmon Gum.

Hollows in gum trees provide important breeding habitat for many species of parrot, such as Rainbow Lorikeets.

Port Lincoln Parrots nest in well-protected nooks in the branches of eucalypts.

Port Lincoln Parrot.

Above, left to right: Crimson Rosella; Sulphur-crested Cockatoo; Eastern Rosella.

Birds roost and nest in eucalypt branches, finding a multitude of uses for the bark,

twigs and branches of gums – from the perfect *perch* to nature's nursery.

Insects, so often unnoticed by humans, use the gradient shades of green to their best advantage.

Merging with the green-threaded leaves is a highly successful survival strategy for this katydid.

Under the cover of darkness, gliders

and possums scurry and *rustle*...

at home in their nocturnal treetop world.

The Leadbeater's Possum licks tree sap.

The Greater Glider eats only eucalypt leaves and buds, making its home in the hollow refuges of tree branches.

54

Left and right: Sunlight burnishes the silhouetted forms and flowers of Australia's most recognisable floral symbol.

The sap and sweet nectar of eucalypt flowers attracts Sugar Gliders by night and birds by day.

Feathertail Gliders dangle from twigs and branches.

Eucalypt leaves are toxic to many animals and are not highly nutritious, yet they are the Koala's only source of food. Koalas have a tolerance for only a few species of eucalypt, making the conservation of these valuable food resources even more important.

More than any other animal, the *Koala*

depends on the eucalypt, for food, for relaxation and for self-preservation.

Koalas also receive all the water they need from gum leaves, which may contain up to 50% water.

Koalas eat more than a kilogram of gum leaves a day, extracting nutrients and filtering out toxic substances.

The Greater Glider is mostly arboreal and rarely ventures out of the treetops where it thrives on eucalypt leaves.

Above: Praying mantis. Right: Rainbow Bee-eater.

Birds, bugs and bees are attracted to eucalypts' *various* forms of blossom and bark.

Above: A mature male Red Kangaroo rests a while under a gum tree. Left: Glen Helen Rock cliffs seen beyond a River Red Gum on the bank of the Fitzroy River, Northern Territory.

In the dusty heart of the continent,

River Red Gums provide shade and *shelter*

for their faunal counterparts, Red Kangaroos.

Branches clutch imploringly at the storm-laden sky,

gather close the night, or pray to the slow-burning sun,

creating so many marvellous moods.

Many Moods

Another day bursts through the crowning glory of gum leaves.

Mottled leaves shine in the early sunlight.

The rays of early morning sun transform

the gum's grey-green attire,

detailing it with *glorious* gold.

Haze blurs the treetops of forests around Booloumba Creek, Queensland.

Mist envelops the forests and lakes, lending a mottled grey *eeriness* to their daytime colours.

Clockwise from top left: Filagree veins in a decaying leaf; mist swirls around trees on Perrys Lookdown in the Blue Mountains, New South Wales; serene waters lap at the roots of Flooded Gum around Myall Lakes, Myall Lakes National Park, New South Wales.

Grass-trees and low-lying scrub fight for sunshine through the eucalypt woodland of the Scenic Rim, northern New South Wales.

A young leaf, tinted in delicate hues of peach and striking fuchsia.

Stringybarks shed their skin in the fern-lined forests of Yarra Ranges National Park, Victoria.

The dying embers of the sun in early *evening*

burnish branches and illuminate bark.

A River Red Gum draws the fading colours from the day's sun.

The late afteroon sun washes this eucalypt grove in drowsy copper, Flinders Ranges, South Australia.

Vivid colour illuminates a gum graveyard in Karijini National Park, Western Australia.

Lurid colour lights the sky and touches,

with a soft *serenity*, the silhouetted

shapes of long-drowned trees.

Ghostly gums stumble and sway over plains near Charleville, Queensland.

Nature's watercolour works of art.

Ghost-grey trunks and wind-whipped branches in the half-light

become screaming *spectres*

bound by rock and root.

An ominous mood pervades this scene in Mt Field National Park, Tasmania.

Eerie portrait of a Snow Gum in Mount Field National Park, Tasmania.

A rainbow scarcely competes with the colour of this gum, standing braced against a storm in Victoria.

Rain on the horizon offers the hope of replenishment for this weary-looking Ghost Gum in the Hamersley Range, Karijini National Park, Western Australia.

Clouds *threaten*

the strong sentine s of the plains with the fear of storms,

yet promise sudden sun-showers of rain.

In still, dark lagoons and billabongs,

the slow rhythms of life *seep* from river to tree

and back again in a languorous ebb...

Gums overhang the muddy waters of a waterhole near Gascoyne Junction, Western Australia.

Reflections dance on the mirrored surface of the Denmark River, Western Australia.

A stately eucalyptus presides over the grassy plains beneath Burt Bluff, MacDonnell Ranges, Northern Territory.

The slim trunks of Ghost Gums snake towards the sky, like slender white arms twirling in a stylised dance.

Ghost Gums were once known as "widow-makers" because they shed their branches in times of drought, killing swagmen camped beneath their heavy boughs.

Left and above: The pure white trunks of these gums add a striking contrast to the Pilbara landscape, Hamersley Range, Western Australia.

Nourished by spring, the eucalypt

flourishes *resplendent*

in deep viridian...

Spindly gums etched against the sunset near Wagin, south-west Western Australia.

Although the DIG tree, made famous by the Burke and Wills' expedition, was vandalised and burnt down in 2002, the tree, near Burketown in Queensland, is nevertheless immortalised in Australian history.

The Australian Labor Party has its roots in the Tree of Knowledge in Barcaldine, Queensland.

Their stories *echo* the hearts, *hopes* and hurts of a nation.

River Gums line the fertile banks of Windjana Gorge on the Lennard River, Napier Range, Western Australia.

The success of the eucalypts' spread throughout the continent is largely due to their simplicity – evident in this image of a single yellow leaf.

Bark and Blossoms

Delicate, sun-like stamens of radiant colour

meet the earthy, time-weathered tones of bark...

Radiating outwards from the woody gumnut, the sticky, scented stamens of gum blossoms are immediately recognisable.

Stringy curls of bark scar the trunks of many trees, giving them an age-old appearance.

Beige bark and *pale*  flowers are

reminiscent of the sun-bleached

landscapes of bush.

Tasmanian Blue Gum, the State's floral emblem.

Colours leap from these gum trees on the Murray River, South Australia, as if an artist in a creative frenzy had daubed them with coats of many colours.

Bright against the blue sky

and green leaves,

are *blooms* in splendid scarlet.

Eucalypt means "well-covered" and refers to the hooded gumnut – the flowers themselves are anything but retiring.

The outermost layer of a gum tree's bark is shed every year, often in large slabs or flakes, as its girth expands, Murray River, Victoria.

Flowering gums add another startling pigment to the palette of the Red Centre, Northern Territory.

Sparse leaves become vibrant *clusters* of leaves and flowers.

Above: A riot of yellow blossoms burst forth from the hard, nut-like buds. Right: Patchy clusters of leaves and flowers match the desolate landscape.

An enduring image of Australia – the gumnut, gum leaf and gum blossom of a Finke River Mallee.

The drooping scarlet blossoms of a Yellow Gum belie the eucalypt's name.

Flowers of many species of gum range from palest cream to a bright yellow.

Many eucalypts have flowers that are almost wattle-like in their softness.

Slender cream branches reach into striking blue space.

Smooth, sleek bark underneath is revealed to, and ravaged by, the elements over time.

The muted colours of the earth

peep from layers *of battle-scarred bark*

and nestle in the knobby clusters of gumnuts.

Yellow Gum Blossom.

Pear-fruited Mallee.

Eucalypt leaves, backlit by the sun.

online

FOR PRODUCTS
www.steveparish.com.au

FOR LIMITED EDITION PRINTS
www.steveparishexhibits.com.au

FOR PHOTOGRAPHY EZINE
www.photographaustralia.com.au

Photography: Steve Parish

Additional photography: Pages 52–3 & 62, Michael Cermak

Front cover: Sydney Opera House and Harbour Bridge. Page 1: Uluru–Kata
Tjuta National Park. Pages 2–3: Snow Gum, Kosciuszko National Park.
Pages 6–7: Flinders Ranges National Park.
Pages 38–9: Mallee Ringneck, South Australia. Pages 66–7: Bunya
Mountains, Queensland. Pages 102–3: Mottlecah bloom.

Text: Karin Cox, SPP & Ted Lewis, SPP

Design: Leanne Nobilio, SPP

Editorial: Michele Perry, SPP & Britt Winter

Prepress by Colour Chiefs Digital Imaging, Brisbane, Australia

Printed in China by PrintPlus Ltd

Produced in Australia at the Steve Parish Publishing Studios